LIVING IN *Pioneer* TIMES

Inspired by days past to become more self-sufficient today

SHIRLEY H. BAKER

KIDS OUTDOOR ZONE
AUSTIN, TEXAS

Living in Pioneer Times:
Inspired by days past to become more self-sufficient today

Kids Outdoor Zone
9508 Chisholm Trail
Austin, Texas 78748
www.KidsOutdoorZone.com

A portion of the proceeds from this book go directly to Kids Outdoor Zone to assist in the mission of "No Kid Left Inside."

PHOTO CREDITS

COVER
Cabin photo © Depositphotos.com/mcornelius (also pg 56);
Wood photo © Depositphotos.com/anterovium
Gingham check and flowers pattern: Footage Firm, Inc.
Water well © Depositphotos.com/rujituk
Honeycomb © duallogic. Image from Bigstock.com.
Butter churn © Depositphotos.com/skopal

INTERIOR
Gingham check, dot pattern, and flowers patterns throughout:
Footage Firm, Inc.
All other images © T. J. Greaney or Shirley Baker unless otherwise labeled.

Cover and interior design by Tamara Dever for TLC Graphics, www.TLCGraphics.com

ISBN: 978-0-9909924-0-0

Printed in the USA

Contents

Recipes

Introduction

For many pioneers, bringing food to the table was a family affair. The father hunted and farmed. The mother prepared and preserved the food. The children were expected to do their share, too, hauling water, feeding animals, setting the table, washing dishes, or caring for younger siblings.

Only essential cooking utensils would have been carried over the trail to the family's new sod house or log cabin dwelling. Their kitchenware might have included these items:

- teakettle
- saucepan
- coffeepot
- coffee grinder
- milk pail
- cake pan
- tin plates
- cups
- utensils
- cast-iron fry pan
- cast-iron Dutch oven
- wooden bowl
- chopping tool
- cutting board

Cooking was plain and sometimes boring because only basic foodstuffs were available, due to lack of money or a large producing garden.

Nothing was wasted. If flour, sugar, cornmeal, or salt came in cloth bags, the fabric would become someone's garment, a bed cover, or, if the house had windows, curtains.

If the family was fortunate enough to have a large log cabin, a second story was used as a children's bedroom and possibly as a storeroom for grain. The children slept on the floor on pallets stuffed with corn shucks, leaves, pine straw, or grass. The parents kept the fireplace below going during the long, cold nights, and the heat rose up through the cracks in the floor to the children above. The parents slept on a trundle bed downstairs. Any small babies slept on the pullout part of the bed. Having the infant nearby made it easy to feed or comfort him or her during the night.

Our ancestors were masters of creativity. If eating utensils were not available, gourds were hollowed out and used as bowls or water dippers. Food was eaten with the fingers until someone carved spoons out of soft wood. Black sweet gum branch tips were frayed and used as toothbrushes. Corn shucks were tied together and used as brooms.

Families shared close quarters and worked hard. There was little time for squabbling. They were thankful for what little they had — especially one another and their faith in God.

This book will give you just a small look at pioneer living. A lot of effort went into what we take for granted today: clean bodies and clothes, nourishment for our bodies, and resting places. There were no grocery stores, no nearby neighbors from whom we could borrow a cup of sugar. Pioneers had to be self-sufficient. In return for their work, they received the tranquility that comes from living close to nature and the satisfaction of surviving through their own efforts.

Adobe Brick

Sand is the filler and clay the binder for adobe brick. Sand is about 25 percent of the mixture, as too little clay makes bricks weak and crumbly, and too much clay makes bricks too dry. Molds may have been made of wood. Usually, 4 x 7½ x

16 inches was the finished size of the brick. Molds could be made in long slabs and then cut, and molds held their shape best if put together with screws, not nails. Molds were immersed in water after each was emptied to prepare for another filling and help bricks to come out of the mold. A good protective covering for adobe walls, both inside and out, is linseed oil, but it was rarely found in pioneer homes.

Three coats of oil were applied if used. Inside walls were treated by applying one part quicklime, six parts cottage cheese, and water to create a smooth-flowing mixture. Of course, other mixtures were used according to the availability of ingredients. Possibly buttermilk paint was used (1 gallon buttermilk to 4½ pounds of white cement). Many pioneers must have lacked any of the above ingredients and had to devise their own wall surfacing. One possibility: grinding their own limestone into a powder and adding water. *See also* dwellings.

A good protective covering for adobe walls is linseed oil.

Baskets

Pine needles, sweet straw, or peeled white oak were used to make baskets. Vines were woven to make a particularly sturdy basket.

CLOCKWISE FROM BOTTOM CENTER:
wool/cotton corder with metal teeth;
Kudzu basket; wood spindle for spinning wool

Brooms

Broomcorn was normally grown by farmers to make brooms. The task of making brooms was done when the weather was bad and outdoor activity was not possible. Birch saplings, white oak, or ash were sometimes used to make a handle. There were many variations on how to make a broom, depending upon the skill of the maker. For a simple broom, lengths of broomcorn (about thirty or so) were gathered, and the seeds were combed out with a currycomb. Twine was wrapped around the stalks, forming a handle. A more sophisticated scrub broom was made by cutting a 5½ x 14-inch piece of wood, drilling twenty-one holes into it, and drawing corn husks through. A handle made of a tree limb was placed in a middle hole that had been bored on the slant.

Brooms were typically made during inclement weather.

© Depositphotos.com/skopal

Butter

Butter was made by skimming the cream from milk and saving until enough was gathered to churn. A churn was mostly made of wood and had a lid with a hole for the *dasher*, which was a long pole with two pieces of wood crossed at the bottom and moved up and down to agitate the cream. When the butter was formed, it was scooped out of the buttermilk and washed repeatedly in a bowl of cold water. It was then salted, kneaded, and molded into various shapes. Butter, milk, and eggs were kept cool by lowering them into the well, since no refrigeration was available.

ABOVE: *Oak butter churn*

Candles

Beeswax burned more slowly than paraffin or tallow and was quite a luxury. Tallow candles had a "fat" smell but were useful and effective. Beeswax or beef or sheep tallow were placed in a can with water and boiled for an hour or so then the wax was skimmed off the top for candles. Candlewicks were made from heavy string that had been soaked for at least twelve hours in a solution of 1 tablespoon of salt plus 2 tablespoons of boric acid in a cup of water. After the string dried, it was braided to form a wick. The string was then put on a wooden dowel and dipped in melted wax over and over until the desired thickness of the candle was met. Then the candles were hung to dry.

Candle mold and string candles.

Compared to tallow or paraffin, beeswax was a true luxury!

Cattails

Cattails were used throughout the year: *Fall*: White seed head for pillow stuffing, insulation, tinder. *Summer*: Yellow flower head, remove husk, boil, eat like corn on cob. Put bag over flowerhead, shake — flour! Leaves may be used to weave baskets *Spring*: Cut close to bulb, peel and eat raw; slice and boil. Resembles cucumbers in taste.

Winter: Underground bulbs, potato substitute. Dried and ground for flour. Boil in salt water, peel outer husk away, use only inner core.

Clothing

Clothing was homemade, sometimes made out of feed sacks. If there was money to buy fabric, wool, flannel, or calico was chosen. It is possible that women sometimes bartered exchanging eggs for fabric.

ABOVE:
Typical dress.

RIGHT:
Spinning wheel

The spinning wheel most likely originated in India between 500 and 1,000 A.D.

Coffee

When coffee beans were not available, other methods were used to satisfy the thirst for coffee.

Breadcrust coffee: Stale crusts of bread were toasted until dark brown, then put into a large pitcher with boiling water poured over to cover. The pitcher was covered and left standing until cool. The "coffee" was then strained and sweetened to taste as desired. The coffee was reheated to boiling when ready to serve. (Today's Postum is a grain powder and still sold in stores.) Stale bread was also used to make French toast and bread pudding.

Sunflower Seed Coffee: Shell sunflower seeds easily by crushing with rolling pin. Drop into vessel of water. Kernels will sink and hulls will float. Reserve seeds for eating. Heat empty sunflower seed hulls in skillet till brown. Put through grain mill. Use 1 teaspoon ground hulls to each cup of water. Steep for three minutes, strain, sweeten with honey if desired.

White oak acorn coffee: Acorns were roasted, pounded, steeped in boiling water, and put through a sieve.

ABOVE: *Coffee grinder and enamel coffee pot.*

Cooking on Wood Stoves & Chuck Wagon Cooking

Cast-iron fry pans were put directly over coals to fry food. A Dutch oven was usually hung over the fire. Ironware was greased before placing food in it. If a kettle was in need of repair, the white of an egg was mixed with some iron filings and lime to make a paste. This was applied to the crack and allowed to set.

Usually, food was put in the Dutch oven in the morning and left on the fire to cook all day. There were no quick-fried meals. Having food cooking slowly on the fire freed the woman of the household to do other chores.

Wood stove ashes could be put in the garden to deter slugs and enrich soil; they were also used to make soap. There was always a supply of wood ashes.

Cooking fish soup in a cast iron pot in the fireplace.

Cracked cast iron pan? Mix egg white, iron filings, and lime.

Dutch Oven Cornbread

2 c. cornmeal
1 tsp. salt
1 c. flour
½ tsp. baking soda
1 c. buttermilk

Cooking in Dutch oven placed directly on fire coals.

Rub lard on sides and bottom of cast-iron skillet or Dutch oven. Sprinkle cornmeal on sides and bottom of chosen pan. Pour in batter. Cook on coals which have been pulled out onto the hearth in a pile. Place coals on lid of Dutch oven and bake 20 minutes.

Chuck Wagon Cooking

A chuck wagon, used on cattle drives, was a kitchen on wheels. Every inch of the wagon was used for storage or workspace. Wood or cow chips were stored underneath. In front was a jockey box for storing tools, and in the rear was a tall box, the width of the wagon, which had a stout lid that lowered for use as a worktable. This box held food and utensils, and drawers were sometimes built in for storage. Under this "chuck" box was stored the heavy iron cookware essential for cooking over a fire. A coffee grinder was often attached to the side of the wagon, as were buckets, bedrolls, and tools. A water barrel with a spigot also was a must.

Chuck wagon without the canvas cover.

The chuckwagon was a predecessor to today's food truck!

Dwellings

Adobe: one story, built in arid areas, made of sun-dried mud, possibly with a tile roof. See also adobe brick.

Fireplaces: were built of brick or stone. Free-standing fireplaces radiated more heat but used more fuel and could be found in only a few homes.

Frame: windows faced south, the north was to the rear, the roof sloped, and windows were small to ward off cold.

Log: built on rock piers, letting air circulate and cool in hot weather. Hewn out of pine, cypress, oak, or tulip poplar trees. *See also: log cabins.*

Outhouse built over hole in ground.

Large tree shading the cabin.

Outhouses: called privys. A deep hole was dug into the ground, and a simple box dwelling was built around it.

Sod: less structured than adobe; mixed with grasses. Similar to making sand castles at the beach. *See also: sod houses.*

Stone: had poor insulating qualities, but were fireproof and required no maintenance.

Trees: shaded and protected homes from wind and were planted wisely.

Stone homes were fireproof, but poorly insulated.

Water wheels: Small ones were sometimes used to let the flow of water grind grain. Dams were sometimes used to fill a small pond for easy access to water. Of course, the first thought was to dig a well in hopes of finding water. A well would have been perhaps thirty feet deep.

Windmills: pumped water to the surface and were often used after 1850 as word of their efficiency spread.

LEFT: *Water well with bucket and roof.*
ABOVE: *A windmill and water barrel.*

Edible Plants and Plant Usage

Basil: Repels insects and most often used to flavor stews and meats.

Cactus leaves: Peeled for pulp and the water they contain.

Ginseng: Mixed with milk, honey, or cornmeal to make a fly poison.

Oil of pennyroyal: Repels fleas.

Peppermint: Discourages mice.

Pines, birch, juniper: inner bark is edible.

Thistle: Leaves and roots are both edible.

Flour

When wheat flour was not available, White Oak acorns were harvested and pounded into flour. In winter, cattail bulbs could be dried and crushed and used as flour.

Food

The most important crop to the pioneer was corn. Three times a day, pioneers boiled, baked, or fried corn, feeding not only family but farm animals as well. The shucks stuffed mattresses and made scrub brooms. Corncobs made pipes and the ashes were used as baking powder. These versatile pickings also found their way to the outhouse as a precursor to toilet paper!

Farm animals sometimes made the journey across the land with pioneer families and were used for milk, eggs, and wool. These animals were too useful to be eaten, so wild game and fish provided necessary meat.

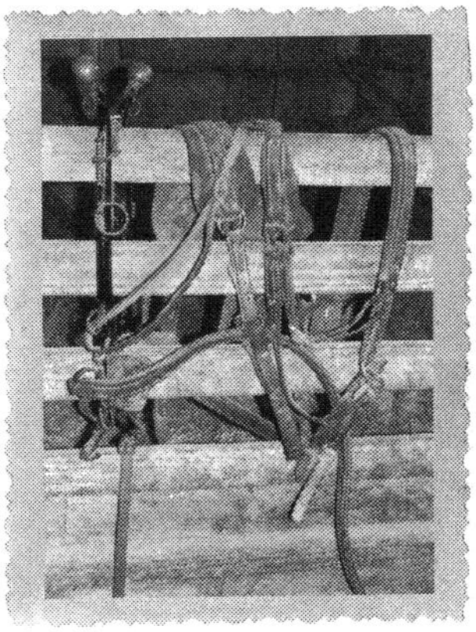 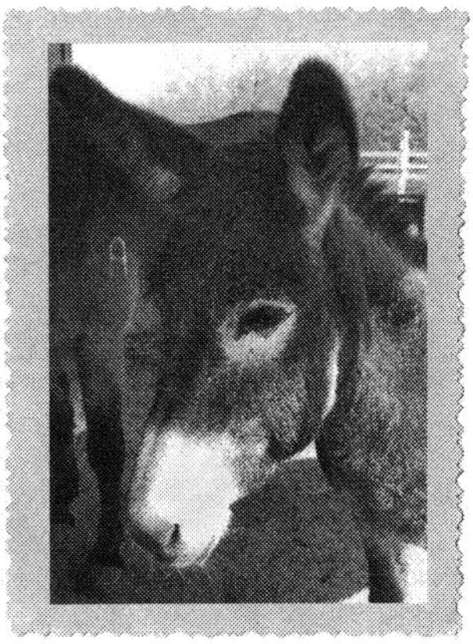

Bridles and halters for animals such as this donkey, used for plowing and harvesting crops like wheat and corn.

Soup and stew were the common meals. Ingredients were thrown in a Dutch oven and hung over the fire to simmer all day while the cook tended to other chores in the cabin or outside in the garden.

Cornhusk dolls were made by little girls, layering the cornhusks and tying with string or a thin strip of the cornhusk.

Gardening

Seeds were precious cargo for the pioneers as they traveled across the prairie. What few seeds they had were planted and nurtured to sustain the family. Without modern-day insecticides, old-time methods were used in the garden:

Planting onions among cabbage plants deterred cabbage moths.

Planting marigolds around beans discouraged bean bettles.

Fireplace ashes sprinkled on gardens added to soil and discouraged some bugs.

Snuff sprinkled on anthills made ants scurry elsewhere.

LEFT: *Cabbage* RIGHT: *Spinach*

Cabbage worms were discouraged by boiling hot peppers in water and pouring over cabbage. Dusting with fireplace ashes also discouraged worms.

Petunias were planted between potato plants to deter potato bugs.

Corn planted in a square, in rows, helped pollination.

When dry beans were picked, they were put in a burlap sack and beaten with a stick. The beans fell out of the pod and were ready to store.

LEFT AND ABOVE: *Planting corn and a garden, and building a house would come first, then animals — they need food, too!*

The family worked together in the garden, which was their only source of food other than any animals or fish they hunted.

When cabins were elevated off the ground by large rocks, chickens could roam underneath and feed on bugs that could otherwise find their way inside the cabin.

Glass

Glass windows were rare, as glass was expensive and had to be shipped in, usually from the East Coast. Glass dishes or tumblers were proud possessions and well cared for, because they were usually handed down from generation to generation.

Kerosene-soaked string was tied around a glass bottle if one was available. The string was lit with a match, and when the fire burned out, the bottle was placed in cold water. The bottle would separate, and the rough edges of the bottle were filed. These were used as bowls, candle holders, or drinking glasses.

Kitchens

Kitchens of the past were a section of the cabin designated by a fireplace, perhaps a window, and shelves or a cabinet for food and cooking-tool storage. A primitive table could be wood planks on sawhorses, with chairs, benches, stools, or tree stumps for seating.

The kitchen was the center of activity, where spinning, weaving, candlemaking, sewing, and reading were done. In the early pioneer days, there were no matches. Fires were started by flint and tinderbox, so keeping a log fire going was essential.

For cooking, skillets had legs and sometimes had long handles to keep safe distance from the hot burning coals. For frying, pans could not be put directly on the coals, because that would put out the fire. One-pot meals simmered all day, hanging over a fire by a crane built into the side of the fireplace.

Eating with a knife was the norm in the Midwest during the first half of the nineteenth century. Knives had a spatula look with a rounded end for placing food in the mouth. Forks came along with two or three tines. Even in those days, the forks were on the left, tines down, and the knife was on the right. A revival of the square "trencher" (a square dish carved from part of a tree) has come to our tables by the makers of dinnerware and truly a delight to see.

Dried beans, peppers, herbs, and utensils.

If the family was large, the children often had to eat standing at the table for lack of chairs.

Lanterns

Early lanterns were bottles filled with kerosene, and rolled rags were used for a wick. Double glass globes then were used: the kerosene filled the bottom, and the wick extended to the globe at the top, which protected the flame from the wind and could be extended for more light as needed.

Mason jars, candle mold, string candles, and oil lamp were prized possessions.

Laundry

Water was a precious commodity and was never wasted. On wash day, water was usually hauled by the children from a stream or a well to a wooden or metal tub. A portion of the water was heated over a fire and added to the tub, or the metal tub could be used to warm water over a fire, but adding water to these large tubs made them extremely heavy. A scrub board and lye soap (made from wood ashes and fat) were used to clean laundry. Clothes were rinsed in a second barrel or in a stream, then hung on a rope strung between trees or placed over bushes to dry.

Another method of doing the laundry was putting a large pot over a fire outside. One would boil the water and lye soap, add the clothes, boil some more, and rinse in a clean water tub or take to the creek for rinsing.

When moving to new land, pioneers would sometimes bring a large barrel of water for laundry and drinking tied to the side of the wagon.

Oak washtub and scrub board.

Notched logs (saddle notch) were popular for log cabins.

Log Cabin

The log cabin concept originated with Scandinavian immigrants. Pioneering Swedes first settled in the New World around 1638. Their settlement was called New Sweden and today is known as Delaware. Cabins were thought to have first appeared in what is now Georgia in the 1700s. A normal size for a cabin was 16 x 16 feet. It would have had one door, and maybe one window covered by animal skin until glass was imported and could be afforded. Perhaps a loft was on one side for children's sleeping quarters, a ladder leading up to the small area. The ceiling was low, which made warming the cabin in the winter easier. In the summer, the mud and grass between logs, called *chinking*, was knocked out in places to let a breeze flow through. In the cooler months, the mud chinking was replaced. *See also* dwellings.

Lye

Wood ashes and water were allowed to stand overnight. When the water was poured off in the morning, it was essential to cover eyes and skin with glasses and clothing, as the very dangerous liquid was caustic lye!

Straw was placed in the bottom of a half wood barrel with a hole on the side near the bottom. The barrel was then filled with wood ashes. This barrel was elevated by placing on rocks so that an enamel pan was placed under the hole in the barrel. The center of the ashes was scooped out to hold about three quarts of water. The water was heated to boiling, and this process was repeated daily until lye started seeping out of the hole. The pioneers were very careful when making lye and soap, and young children were kept away during this process.

Medicinal Old-Tyme Remedies

Arthritis: Honey and vinegar and whiskey; A magnet to draw pain out of body; Pine needle tea to relieve aches and pains.

Asthma: Keep chihuahua dog in dwelling; Honey and lemon juice and whiskey.

Athlete's Foot: Garlic and olive oil.

Bleeding: Spider web was matted on wound.

Bruises: Arnica gel.

Burns: Aloe vera leaves. Sunburn could be eased with wet black tea bag, lavender oil, vinegar water, or witch hazel.

A typical size for a log cabin was 16 x 16 feet with one door and maybe a window.

Raw honey had many uses including treating cuts and colds, sore throats, and as a sedative.

Cuts: Aloe vera, honey.

Chest congestion: Raw honey eaten by the spoonful. A flannel rag soaked in lard, turpentine, and kerosene was put on the chest. Pioneers also used a mustard plaster which was ¼ cup dry mustard, ¼ cup flour, 3 tablespoons molasses, and lard. This was mixed well. The chest was covered with a piece of flannel that had been dipped in warm water and wrung out. Then the mustard plaster was spread on the cloth, and the cloth side was placed on the patient's chest for no longer than 15 minutes.

Colds: Honey and whiskey. Onions, roasted. Honey and vinegar. Ginger tea.

Colic: Warm towels placed on stomach, or hot water bottle wrapped in a towel. Infant was placed on stomach over parent's legs. Ginger tea was also used for colic.

Constipation: Teaspoonful of powdered sulphur and molasses twice a day.

Cough: Whiskey and sugar. Honey and vinegar.

Dandruff: Yogurt rubbed into scalp, left on one hour, washed out. A vinegar rinse was then used.

Depression: Cup of peppermint tea.

Diarrhea: Blackberry juice; Prickly pear cactus, split, and roasted pulp was then eaten (Indian remedy).

Dry skin: Lanolin or buttermilk applied to skin.

Earache: Juice from a sweet shrub bloom placed in ear. Castor oil drops, warm. Mineral oil drops, warm.

Energy: Cider vinegar

A mustard plaster was applied with a damp, warm cloth to help clear chest congestion.

Everything salve: One pound resin plus ¼cup mutton tallow, half as much beeswax, and ¼ ounce camphor gum. Put into old kettle, boil until dissolved, stir with stick. Take ¼ pail warm water, pour it in, and stir until white and brittle. Grease hands and keep wet to prevent sticking. Wet table, roll out salve, cut with knife into cubes, keep in cool place.

Eye inflammation: Cloth soaked in sassafras tea and placed on closed eye gives comfort.

Fever: Tea made from bark of dogwood tree. Oak bark also used.

Foot soreness: Lavender oil drops and warm water.

Gall bladder: Corn whiskey and black draught.

Gas in stomach: Sage tea, a cup as needed.

Headache: Basil tea. Application of hot water to the feet, then back of neck, as needed.

Hiccups: Peanut butter by the teaspoon.

Insect Bites: Rub with an onion. Rub with Vick's® VapoRub.

Jaundice: Cherry bark tea for liver ailments

Muscle ache: 1 cup witch hazel and ¼ cup rubbing alcohol.

A spoonful of peanut butter to cure hiccups.

Nausea: Cloves, Coke syrup, ginger.

Nosebleed: Hold nickel in mouth under upper lip. Folded paper also used.

Poison Ivy: Rub wood ashes on rash. Also, a poultice made from bruised leaves of the nightshade and cream. Tar soap with mixture of jewelweed is most soothing. Drinking milk and ingesting vitamin C are also recommended. Tar soap (wet, rubbed on skin, and left on skin), is the most drying.

Rheumatism: Sulphur baths, sulphur tea.

Scratches and scrapes: Apply honey.

Seasickness: Teaspoon of red pepper, mixed with molasses, taken in one dose.

Sedative: Honey.

Sleep: Cider vinegar.

Sliver, thorn, or boil: 1 tablespoon stale bread, ½ cup milk, boil. Add little glycerine and sweet oil to make pliable. Apply and replace as needed.

Snake bite: Apply freshly killed animal or chicken (an old miner's tale). Mud packs applied to wound; turpentine and gunpowder, kerosene and salt, tobacco juice. (Remember, doctors and transportation were scarce.)

Sore throat: Honey and lemon juice; A cup of warm water and 2 tablespoons vinegar, gargled.

Sprain: Apply cayenne poultice or lavender oil.

Teeth: Clean with twig toothbrush, made from black-gum tree or sweet birch, which has been frayed at the end. Salt or baking soda were sometimes used as a toothpaste.

Toothache: Oil of cloves.

Warts: Rub castor oil on warts for several weeks. Touch with acetic acid (vinegar) often during the day.

A few drops of lavendar oil in warm water can relieve sore feet.

Pectin

Pectin was made from fruit and is a gelling agent used mainly in making jams and jellies.

Homemade Pectin

Select small, green, immature apples, wash thoroughly, removing any spots or bruises, and cut into thin slices. Put in pan; add 2 cups water for each pound of apples. Boil 15 minutes. Strain through cheesecloth. Put pulp back into vessel, adding required measure of water.

Cook again 15 minutes, over low heat. Let stand for 10 minutes. Strain juice with pulp through cheesecloth. When cool, press remaining juice from it. To keep for future use: Heat jelly stock to boiling point; pour into hot canning jars. Seal. Invert and allow to cool. Keep cool.

ABOVE: *Apple barrel*

Grape Jelly

21/2 c. grape juice   1/2 c. honey
2 c. homemade pectin

Blend and boil all ingredients for ten minutes. Pour into hot jelly jars. Seal. Keep cool.

Pests

Ants: Save cucumber peelings and mix with salt. Place wherever ants run. To keep them off of the table, place table legs into cups partially filled with water.

Water poured into a tin cup kept ants off the table.

Those before us used water, borax, or leaves as pest control. Treatments today have injured the ecosystem by adding pesticides to waterways and the ground.

Flies: Mix castor oil and melted resin. Spread mixture on slick paper. To keep out of house, mix bay leaf pieces, ground cloves, broken eucalyptus leaves, and clover. Put blend in mesh bags and hang just inside doors.

Roaches: Place shallow jar lids containing borax in dark corners of house. Replace once a week. Keep away from pets and children and from around food. A mixture of borax and evaporated milk is also effective.

Mice: Mix cornmeal and cement. Place in shallow container where mice run.

Squirrels: To discourage squirrels from nesting in a home, holes that they make by chewing were stuffed with wire or covered with tin.

Quilts

Fabric was precious, so scraps of material were always cut from worn-out clothing, feed sacks, or bedding. It sometimes became necessary to crazy quilt because of different fabrics, sizes, and amounts. Quilts kept people warm, were used as clothing or bedding, and were treasured gifts for newlyweds. Women spent

hours socializing while quilting, making every minute together count. Cotton or wool was used as filler if available; plain cotton was preferred for backing. The backing could also be made of flannel, adding warmth to the quilt.

A quilt frame was usually hung from the ceiling and brought down to work on. When finished for the day, the rack with the attached quilt was raised back up to the ceiling.

Soap

Soap was usually made outdoors, over a fire, using a cast-iron kettle suspended from a birch sapling, wigwamlike frame. Late autumn was the season for making soap, as that was the time when butchering took place.

Bars of soap. The lighter ones are the natural color; the darker ones are colored with beet juice.

Following are a few recipes for making soap like the pioneers did. Small children must never be near when soap is being made, due to the caustic nature of the lye.

While many soaps are pure, some are scented, some medicinal. The art of making soap has become a big business by creative people in today's world.

Pioneer-Style Soap

Following is a basic recipe for making soap like the pioneers. Small children must never be near when soap is being made due to the caustic nature of the lye.

To make lard: One cup melted beef tallow, cut into small pieces and heated over low flame. Strain through cheesecloth to remove cracklings.

---

1 can lye	1 tbsp. borax
1 qt. water	5 lbs. lard
¼ c. kerosene	1 tsp. bleach

1 tbsp. ammonia

Put one quart water in earthen crock or enamel container. Add lye to the water and stir so the lye does not settle to the bottom. Be careful that you do not inhale fumes.

Cool lye and lard separately. Add kerosene, ammonia, borax, and bleach to lard. Pour lye into lard slowly and evenly, stirring slowly until mixture has thickness of heavy sour cream. Line cardboard box with heavy paper; pour soap mixture into box. Let set a day. Cut into bars. Let age two weeks before using. Use only earthen crock or enamelware, or a cast-iron wash pot for making and cooking soap.

Soap Scents

Pine: Soak pine needles in water; use in recipe above.

Flowers: Spread lard out. Put flowers in for 24 hours so the soap can absorb the fragrance.

Sod Houses

Surrounded by soil, sod houses were usually the first dwelling for pioneers if they didn't live in their canvas-covered wagons. In dry weather, dirt sifted down from the sod-house ceiling; in wet weather, mud oozed down. An industrious woman would place fabric on posts over the bed to catch dirt or the bugs that inhabited the sod (a pioneer four-poster bed!).

The floors were usually uneven dirt, swept several times a day. Field mice ran everywhere, so food had to be carefully stored. *See also* dwellings.

Sod houses were made of soil and grass and easy to build by the whole family. Thick walls helped keep out cold in winter. They were also less likely to burn down than a log home.

Sourdough Starter

Yeast was not always available, so the reliable sourdough starter was kept and handed down for generations and was the pride of cooks in a home or chuck wagon.

Sourdough

The recipe to begin:

4 c. flour	4 c. warm water	3 tbsp. sugar

Mix together, let stand two days until it sours and bubbles, keeping covered and in a warm place.

Sponge

1 c. starter	2 c. flour	1 c. water
1 tsp. yeast	1 tbsp. sugar	

Blend above ingredients and let stand overnight, covered in a warm place.

Bread

Add to 1 cup of sponge: 1 cup water, 1 teaspoon yeast, 3 teaspoons salt, and enough flour to make soft dough. Knead and let rise until double. Shape into 2-3 loaves. Let double in size. Bake at 375 degrees for 30-45 minutes, until loaf sounds hollow when tapping.

Redo Starter

Add 1 cup flour, 1 cup water, 1 teaspoon sugar. Shake and let stand 6 hours in cool place until ready to use again.

Stains

Bars of lye soap were used as laundry soap. Stains were rubbed with bar, then laundered in the usual manner. Then the towels or clothing were laid out on grass to dry, to absorb chlorophyl, and to whiten.

Useful stain: Walnut husks soaked in water were used for wood staining. Cloth dye or furniture staining also came from plants and berries.

Placing walnut husks around peach trees also kept out borers.

Storage of Food

Underground storage or concrete blockhouses above the ground were used to keep food over the winter. An underground storm cellar was covered with mounds of dirt. An air vent of some kind was on the top, and a gravel floor was desired. The cellar would usually have shelves for storage and stacking. Smoke houses were also used for storage of food when the smoking process was completed.

Potatoes and apples (if available) were layered in sawdust in a barrel to keep over the winter.

Cellars were used for storing food and also for shelter in bad weather.

Dry Green Beans

Leather breeches (dried green beans).

Young beans could be strung together with strong thread and a large needle; this was one way of preserving/drying green beans. They were then hung in warm air (not the sun). Once dried, they were known as "leather breeches." The beans were reconstituted when needed by placing in water and letting stand overnight. They were drained in the morning and replaced with fresh water. A small piece of ham in the pot with the beans was put on the fire to simmer all day.

Keeping Potatoes

Holes were dug below the frost line. The potatoes were placed in the holes and covered intermittently with straw, dirt, and possibly a piece of tin.

Preserving Food

Mason jars (if available) and hot water bath were used for acid foods such as tomatoes and pickles. One must remember that glass was at a premium, especially canning jars.

Cookies were kept soft by storing with a slice of bread in a jar.

For apple storage, a barrel was put halfway in the ground, covered with a leaf-filled burlap sack, and then covered with dirt and rocks.

Salting, drying, or smoking were the means of preserving meat.

Cucumbers, onions, cabbages, and sometimes meat were pickled.

Brown and white sugar

Sugar

If one was fortunate enough to buy sweeteners, both molasses and brown and white sugar came in loaves. Because loaf sugar was expensive, sugar was often made from beets. It was not pure white, but is nutritionally superior to commercial sugar.

Beet Sugar

Scrub beets well and chop into small pieces. Cook in water to extract juice; then strain juice. Cook down, resulting in a syrup. Continue cooking until it crystalizes.

Can you imagine using sugar sparingly, if at all? Women rationed their precious commodity carefully.

Honey can be substitued for sugar: 1 cup sugar = ¾ cup honey. Use ¼ cup less liquid in recipes. (Only half of the sugar in a cake recipe can be substituted with honey.)

Peach cobbler was a favorite dessert from chuck wagons and the cabin hearth. Today we can use the slow cooker to get a similar result.

Shirley's Peach Cobbler

6 oz. dark brown sugar

3 1/2 oz. rolled oats

4 oz. flour

1/2 tsp. baking powder

1/2 tsp. ground allspice

1/2 tsp. grated nutmeg

1/4 tsp. salt

1/4 c. butter at room temperature plus extra for the cooker

20 oz. peach slices, canned or frozen.

Combine sugar, oats, flour, baking powder, allspice, nutmeg and salt in a large bowl. Add butter and work in the dry ingredients until a crumbly texture is formed. Fold in the peach slices.

Butter the bottom and sides of a 3 quart slow cooker. Add the mixture to the slow cooker and cook on low for 3 to 3 ½ hours. Serve immediately with heavy cream or ice cream.

Tea

Apple: Dried apple peels steeped in hot water.
Mint: 1 teaspoon dried mint leaves to 1 cup boiling water. Steep for a few minutes.
Parsley: Fresh parsley steeped in hot water.
Strawberry: Collect green leaves, 2 handfuls into a teapot. Add boiling water. Sweeten with honey and serve.

Tools Needed

A post-hole digger and fence tools were always needed.

Hammers and chisels.

Brooms made of split saplings and broomcorn or straw.

Handles for tools were made from dogwood, White Oak, sourwood.

Shovels, ax, cultivator, hoe, wheelbarrow.

Cultivator.

Could you use a regular saw, a push mower or non-power tool for your chores? Be thankful that today we can use an electric saw, gas trimmer, or power mower!

TOP: *Attach a mule to the long piece of wood, have him walk around the treestump, and he helps grind sorghum.*

LEFT: *A wheelbarrow with a metal tub.*

BELOW: *A wheelbarrow made of wood.*

RIGHT: *Three very useful tools — a froe or frow for making wood shingles, an adze for cleaning and smoothing timbers or planks, and an ax.*

Toothpaste

Honey and ground charcoal were used by Ben Franklin. Bicarbonate of soda paste was often used. Toothbrushes were sometimes frayed twigs.

Turpentine

Pine trees were slashed. A bucket or cup was placed under the slit to catch pitch as it oozed out. This was boiled down to make turpentine. The pitch could also be applied to the fur of a possum, then set on fire. The fur and the fat were burned off to facilitate the cooking of possum for food.

Making turpentine and soap were both very dangerous so children were kept far away when these processes were underway.

Traveling

Going to a neighbor's or a nearby town was quite an event. The family would place their short-legged chairs in the buckboard, hook up a horse or mule, and depart on a family adventure. Chairs had short legs because tall ones would topple over on the rough roads and paths. Also, since furniture was in short supply, having your own chair was thoughtful and convenient when visiting others.

Utensils

The early settlers ate from wooden plates (called square trenchers) and drank from wooden cups that they carved and chiseled out of available wood. They used tin cups if they were fortunate. Rolling pins were carved out of one piece of wood. Bowls were carved out of a log and used for mixing food, and an oblong one was called a *canoe*. Even spoons and forks were carved out of wood.

Vinegar

White wine vinegar: Crush 1 pound raisins, add ½ gallon distilled water, and leave covered with cheesecloth. Let stand in warm place. This will turn into vinegar in about a month.

Fruit vinegar: Boil fruit in equal amounts of water. Press through cheesecloth. To each quart of fruit juice, add ¼ cup yeast. Allow to stand in jar tilted to admit air. Keep in warm spot. Let boiled juice stand for two or three days to ferment before straining. Add yeast after fermented liquor has been removed from the fruit pulp.

Waste Food

Recycling by composting for the garden was a natural thing for pioneers to do. Seeds from fruit or vegetables were always saved for next year's crop.

Water

The type of foliage growing in water is often an indicator of the depth or salinity of the water.

Cattails: water near surface.

Pickleweed: indicates salty water.

Saltbrush: indicates water feet below surface.

Mesquite: water 10–50 feet below surface.

Reeds: water close to surface.

Black greasewood: water 10–40 feet below surface.

Rabbit brush: water 15 feet below surface.

Elderberry shrubs: water 10 feet below surface.

Saltbrush can be easily mistaken for a poisonous plant.

Wild Game

In pioneer times, meat on the family table was usually wild game. Wild meats are good to eat and are frequently higher in protein and vitamins and lower in fat than meats of domestic animals.

To "field dress" means to open the body cavity and remove the internal organs. Any knife with a sharp blade will do. Piercing a small hole through the hide and body wall where the breastbone ends must be done carefully so as not to puncture internal organs. Then the abdomen is slit open. It is best to watch a skillful person do this procedure, as it is difficult for the first few times to identify the organs to cut around and to dispose of.

The diaphragm is to be cut from the body wall; it is the sheet of muscle that separates heart and lungs from other internal organs. It should be cut close to the body wall, following the ribs down one side, across the backbone, and up along the ribs on the other side to the starting point. To free the heart and lungs, reach forward until the windpipe and esophagus are felt, cutting them off as far forward as possible. Free intestines by cutting around the anus and pulling them through to the inside. This will be easier if you slit the pelvis with a heavy knife. When field dressing large animals, a small ax is used.

Hooks for hanging meat in a smokehouse.

The internal organs can now be removed in their entirety by rolling them out onto the ground. Wipe the carcass clean and dry inside. Do not use water unless the organs were punctured by shot or knife. Separate heart, liver, and kidneys and put them in a bag or skewer on a sharp stick for easy carrying. Hang the carcass to cool quickly. Prop the body cavity open with a stick. During warm weather, get to a cool spot as soon as possible.

© Marcischauer | Dreamstime.com

Antlers were used for hanging up hats, clothes, etc.

The best deer meat is in the hindquarters and along the backbone. To skin a deer, hang by antlers or neck with a stout rope. Cut the hide around the neck and down the belly. Peel hide down until enough is free to place a rounded stone under it. Tie a rope around the bulge and tie the other end to a tree trunk or heavy object while a partner uses a knife to free the hide from the meat. Preserving the hide may be done by removing all meat and fat. Salt heavily and dry in a shady place. One can never use too much salt. Hides were used for clothing, hats, bed covers, shoes, rugs, or even blankets.

Smaller animals may be cleaned in a similar manner, but hide can be removed by slitting near hindquarters, pulling down, placing under foot, and peeling.

Catfish peeling is done easily when nailed to board attached to a tree. Place fish on the ground and hit in head with hammer to stun him. Hang by mouth on nail; cut around the body just behind the head and just below skin. Grasp skin on either side with pliers and pull down, peeling fish. To finish cleaning fish, slit the body wall down to the vent and remove the internal organs. Cut off the head. Rinse and cool at once.

Today, fishermen clean their fish at sinks provided at marinas; keeping cool is a definite must.

Once again, with any animal, fowl or fish, never puncture internal organs. And always be careful to clean and sanitize tools and hands or anything that touches meat. One cannot be too careful with regard to bacteria and infections. If any organ looks tainted or suspicious, destroy the animal. Play it safe!

Wine, Muscadine

Pioneers in the southern United States made wine from the muscadine species of grapes. The vines thrive in summer heat and humidity. It produces dark fruit with a tough skin and is full of nutrients.

Muscadine Wine

MAKES 4 GALLONS

½ bushel grapes

12 ½ lbs. sugar

© Bryan Pollard | Dreamstime.com

Mash grapes and add 2 ½ lbs. sugar.
Ferment one week. Strain.
Add 10 lbs. sugar. Let work 10 days.

Wood Stoves

REPAIR AND MAINTENANCE

Cast iron stoves were top on the list of a woman's possessions to carry West. They learned its temperature by holding their hand in the oven. A certain number of seconds meant it was ready for baking the bread.

Oven putty: Paste may be purchased in cans and used to repair small cracks.

Stove cement: For cast-iron stoves, mix egg white and hydrated lime into a paste. Add iron filings. Mix well, putty onto stove to repair, let dry. Gets harder as stove is used.

Wood-ash stove repair: Mix equal parts of fine powdered wood ashes, powdered fire clay, salt. Add water to make paste. Needs replacing occasionally as it dries and cracks.

Wood Uses

Oakwood was used for sustaining heat or cooking. Other soft wood trees were split for kindling to start fires. Through trial and error, the pioneers found which trees were good for which purpose — strength or simply useful.

Black gum: twig toothbrushes.

Cypress: logs for cabin.

Dogwood: handles.

Hickory: firewood, smoke houses, handles.

Oak: logs for cabin.

Pine: logs for cabin.

Oakwood was used for sustaining heat or cooking. Other soft wood trees were split for kindling to start fires.

Sourwood: tool handles.

Tulip poplar: logs for cabin.

White Oak: baskets, chair bottoms, fences, tools, barrels, furniture.

ABOVE: *Oak fence and barn, making use of available wood nearby.*

Inside the fence a farmer would plant crops hoping to keep large farm animals from devouring. Fences also kept pigs from wandering and routing out mama's garden or flowers.

Yeast

Yeast is a leavening agent used to make bread rise. Because it was difficult to control yeast cultures, pioneers would reuse doughs and starters to leaven future batches of bread.

YEAST

Steep 1 quart fresh, well-washed peach leaves in 3 cups of boiling water for fifteen minutes. Drain, adding enough water to make 3 cups. Bake 3 medium potatoes, peel, put through sieve. Scald ½ cup cornmeal in 1 cup water until it boils and thickens. Stir to prevent lumps. Put all ingredients in bowl with 2 teaspoons of salt and 3 tablespoons of sugar. Cover and allow to ferment in warm place for 24 hours, stirring every 3 hours. Pour into glass jar; keep cold. Stir down several times until foaming ceases. When ½ inch clear liquid rises to surface, it will be ready for use. Stir thoroughly each time you use it.

When starter is reduced to 1 cup, add 3 cups water, 3 baked potatoes, scalded cornmeal, salt, and sugar as the first time. Leave in warm spot. In about 7 hours it will become active.

Starter improves with age but should be used once a week. If not, stir every couple of days, adding 1 teaspoon sugar.

To dry yeast, sterilize 2 quarts of cornmeal in low oven for an hour. Mix into starter. Spread into flat pans ½ inch thick. When set, cut into 1½-inch squares. Move apart to dry and harden. Wrap cakes. Store in cool place. Will keep for a year.

Yeast Starter

Starter made from dry yeast in previous recipe.

© Depositphotos.com/ jochenschneider

Fresh and dried yeast

Mix 1 cake yeast, ½ cup warm water, ½ teaspoon ginger, 1 teaspoon sugar. Keep covered until white foam is on top. Stir in ½ cup water, ½ cup flour, 1 teaspoon sugar. Allow to foam, stirring often. Pour into jar and cool. Put lid on loosely until foaming stops. When ½ inch clear liquid has risen to surface, starter is ready to use.

Baking with yeast is a complicated method to making fresh bread but well worth the time and effort.

Glossary

adobe: house of sun-dried dirt and straw bricks.

broomcorn: a type of corn with stiff branches and flower clusters, often tied together and used as a broom.

chinking: mix of mud, grass, and water placed between logs.

compost: leaves, grass, and vegetable matter left in a pile to rot to form fertilizer and mulch.

Dutch oven: black cast-iron pot with handle and cover that is hung over a fire to slow-cook food.

leather breeches: green beans strung and dried for future use.

lye: caustic substance made from wood ashes and water, used to make soap.

log cabin: cut trees placed together and sealed with chinking, with few or no windows and doors.

pallet: fabric filled with straw, grass, and corn shucks, and used as mattress on floor for sleeping.

pectin: jelly stock made from green apples used to make jelly.

Chinking was removed in places during warm weather to let the air flow through the cabin, then replaced in the fall, mostly by the children.

sod house: dirt and grass used with water to build a house like today's sand castles at the beach.

sourdough: starter mixture of sugar, flour, and water to make biscuits and bread. May also make a starter out of potato water.

tallow: white fat from sheep and cows used to make soap and candles.

About the Author

Shirley Baker is a staff writer for *Country Line Magazine* for their homespun section covering fashion, delicious recipes and "Della's Helpful Home Hints." She is also the author of the cookbook *For Goodness Sake*, has transcribed over 50 books for various authors, and is currently working on a novel, as yet untitled, about a young girl and her adventures traveling west on the wagon trail.

Living in Pioneer Times was inspired from her years of baking, sewing, quilting, and simple remedies and stories passed down from ancestors traveling west, who accomplished daily chores in harsh conditions.

Shirley Baker is the mother of five, grandmother of nine, and great-grandmother of six. She lives in Georgia with her husband and their beloved Sheltie, Beauregard (Beau).

A Recipe for The Good Life

AUTHOR UNKNOWN

A heaping cup of kindness
Two cups of love and caring
One cup of understanding
One cup of joyful sharing

A level cup of patience
One cup of thoughtful insight
One cup of gracious listening
One cup of sweet forgiveness

Mix ingredients together
Toss in smiles and laughter
Serve to everyone you know
With love forever after.

Made in the USA
Monee, IL
05 November 2026

67381729R00039